THE ELECTRO-MOTIVE TYPE TURBOCHARGER

THEORY, COMPONENTS AND TROUBLESHOOTING

COPYRIGHT 2009
Technical Training Consultants Inc.
801 Warrenville Rd. Suite 222
Lisle, Illinois 60532

ISBN 0-9842998-2-3

SECTION I

INTRODUCTION

All internal combustion engines produce exhaust gases that contain heat energy. Often - as the case of automobiles - these exhaust gases are simply blown out through the "exhaust system". The energy in these exhaust gases is wasted.

These exhaust gases do not need to be wasted. They may be harnessed and utilized to help the engine run more efficiently. One device capable of harnessing and utilizing engine exhaust gases is called a turbocharger.

A turbocharger supplies an engine with air for combustion. It uses the heat energy of an engine's exhaust gases to drive and air pump. This air pump, in turn, supplies the engine with combustible air.

How do turbochargers harness and utilize engine exhaust gases?

All turbochargers have a "rotating assembly". This assembly has a shaft, and on both ends of the shaft are mounted fans or sets of blades. One of these fans is called the turbine; the other is called the compressor or "impeller".

As exhaust gases pass the turbine, their force rotates the turbine, just as air rotates the shaft to which it is attached. As this shaft rotates, so does the impeller at the other end. This impeller rotates just like a window fan in your home

However, unlike an electrical powered window fan, a turbocharger's compressor or impeller is ultimately powered by an engine's exhaust gas. If you increase either the amount or temperature of the exhaust gas coming from the engine, then the whole rotating assembly - turbine, shaft, and impeller - will rotate faster.

As the impeller rotates faster, it blows more air from is blades, just as an electrical fan will discharge more air when switched from "low" to "high". You might call an impeller a type of "centrifugal pump".

Not all engines have air pumps driven by exhaust gases. Another kind of air pump - called a supercharger or blower - provides an engine with combustible air, but is mechanically driven.

A typical blower consists of two 3-lobed helical rotors encased in a housing. The rotors are driven by a gear train or cogged belt connected to the engine's crankshaft. The energy required to run this type of air pump is thus provided by the same engine that the pump is supplying with air! This situation - in which the blower draws its power from the engine - is called a "parasitic load".

UNDERSTANDING THE ELECTRO-MOTIVE TURBOCHARGER

The EMD turbocharger is a unique type of turbocharger. It has **BOTH** an exhaust - gas **AND** gear-driven turbine whe

Why the unusual design? Because the EMD diesel engine is a two cycle engine which must receive a supply of forced induction air for ALL phases of its operation, **INCLUDING THE STARTING PHASE**. If no forced air is fed into cylinders, no combustion will occur.

Of course, since conventional turbochargers require exhaust gas energy to turn their rotating assembly, they will **NOT** pump air into the engine while the engine's starting motor is cranked.

By adding a gear drive, the EMD turbine wheel uses engine horsepower to drive the compressor and pump air during the starting phase. While starting, the Turbocharger acts as a supercharger.

But once the engine is started and operating at normal speed and load ranges, the exhaust gases flow through the turbine, allowing it to function as a true turbocharger.

SECTION II

THEORY OF OPERATION

How does turbocharged air enter an engine for combustion?

EMD manufacturers its cylinder liners with a row of ports through the cylinder wall, approximately half way up the bore. The ports act as intake valves for the cylinder.

When a cylinder's piston is near the bottom of its bore, air can enter through the ports. Combustible air can now be blown in to charge in the cylinder.

How is this air blown into the cylinders?

Prior to 1959 EMD only used a gear-driven blower. This blower is driven by the crankshaft via a connecting gear train on the engine's rear end. When the crankshaft turns during start up, the two 3-lobed helical rotors in the blower rotate. They drew air into the top of the blower. This air is pumped between the lobes and the blower housing. The blower continues to pump air through the entire speed range of the engine.

However, as engine speed and load increase, so does the engine's need for air. A problem now arises; as the engine's demand for air increases, so does the power needed to turn the blower's rotors, since the blower is a "parasite" working off the engine's power. Eventually the blower's power requirements could exceed the engine's safe operational capabilities.

Of course, the EMD engine is designed so that the blower's power demands will NOT exceed the safe operational capabilities of the engine. With a turbocharger, as the engine's power demand increases, the engine's exhaust gases can now propel the turbocharger's turbine wheel, supplying the additional air needed to satisfy the engine's power requirements. The EMD turbo is equipped with a clutch assembly which allows the turbine rotor to spin without relying on the connecting gear train. Thus, the engine can now receive enough air without being dependent upon the engine's gear train.

SECTION III

COMPONENT FAMILIARIZATION

1. The Turbocharger Nameplate

The nameplate includes a model number. To discover exactly the type of turbocharger required on your engine, consult and EMD parts catalog. For example, catalog #300 provides a turbo application list on Parts List #174.

The nameplate also includes a serial number, which indicated the date, production sequence number, and assembly location of the turbo. For example:

88-A1-1005

88 = year

A = month (A=January, B=February, etc. though the letter I is not used, so December begins with M)

1 = type (1 = new unit; 2 = a repaired or returned machine; 3 =a unit exchange - UTEX - turb

1 = plant (1 = LaGrange, IL)

005 = sequence (005 = the 5th turbo built at that plant during the month indicated)

Finally, the nameplate has an identification code, which indicates the turbo model, gear ratio, and that model's most significant change or revision from the original design. For example:

3E 17.9 R

3E = model (E = 645 engine turbo; T = 567 engine; G = 710 engine)

17.9 = gear ratio (measured in ratio to crankshaft speed; thus, a 17.9 indicates that the turbo runs at 17.9 times the crankshaft's speed. 16.8:1 and 18.1 are also common gear ratios.)

R = revision (Indicated the latest revision applied to the turbo)

2.The Doweling Assembly

The doweling assembly encases all of the turbo's internal components. It includes six iron castings which are aligned to one another by dowels and held together by various threaded fasteners. The alignment of these parts is critical, and the bore through which the turbine wheel passes can be no greater than .0005 total indicator run out.

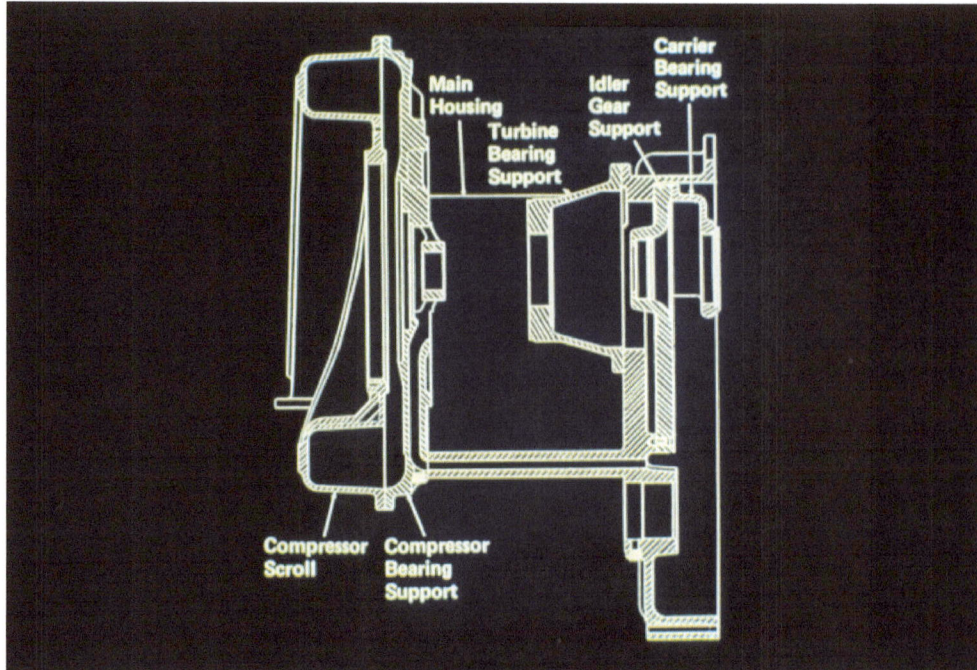

To ensure proper alignment of the six castings, EMD aligned the six pieces and then dowels them together. EMD then stamps the pieces with "doweling numbers" which identify them as a matched set. If one of these six pieces becomes damaged during the life of the turbo, a new part must be aligned to the remaining five components. This new part will then receive a matching doweling number to identify it as a part of the original set.

Each of the six pieced has its own name and function:

1. The compressor scroll forms the "scroll" through which compressor air flows from the turbine wheel to the engine.

2.The compressor bearing support provides a location points for the turbine wheel compressor - end support bearing. The bearing support also forms the rear half of the air scroll.

3. The turbine bearing support provides a location for the rotating assembly turbine - end support bearing. It also contains the planetary gear system on all turbos and the overrunning clutch on 567 and 645 turbos.

4. The main housing is the central component to which all other attach.

5. The idler gear support attaches to the "back" of the turbo and contains various threaded holes for the attachment of the external gears which connect the rotating assembly to the engine gear train.

3. The Main Housing "Cradle" Gasket Area

On the EMD turbocharger, the surface between the main housing and the compressor bearing support - also known as the "cradle" - needs sealing. The specific area which requires sealing is the oil drainage passage at the bottom of the cradle.

Originally, to seal this passage, EMD used a paper-type gasket covering the bottom one third of the cradle. On each side of the paper gasket was ANOTHER gasket made of metal shim stock. These two metal gaskets were not required for the seal; rather, when the paper gasket was installed, the metal gaskets helped maintain parallelism between the two doweling components.

However, because this three piece conventional gasket did not fare well under heat and vibration, EMD designed a more durable seal. In the late 1970's, the company changed the turbo's main housing oil drain from an oval shape to a double round hole configuration with counterbores for o-rings. The o-ring design required NO gaskets between the main housing and compressor bearing support.

What about the older turbo castings made with oval-shaped openings? To improve their sealing capacity, EMD developed a new seal known as the "Parker Seal". This replacement seal is equipped with an oval-shaped o-ring on each side of a metal plate.The seal is then retained when the two doweling components are bolted together.

Parker Seal

6

4. The Turbine Wheel

The heart of any turbocharger is the turbine wheel, also called the "rotating assembly". It includes a shaft with both the turbine's blades (or exhaust fan) and the impeller (or air compressor fan). Near each end, the shaft is supported by two bearings. The bearing nearest the impeller is called the compressor bearing; the bearing nearest the turbine blades is called the turbine bearing.

On the EMD turbo, the extreme end of the shaft (near the turbine blades) also has a small gear. This gear works with other gears to connect the turbine wheel to the crankshaft, for reasons discussed in our introduction.

To minimize vibrations at high rotational speeds, the rotating assembly MUST be well balanced. Let us now examine the components of the turbine wheel nor "Rotating assembly". We will begin at the impeller end

1. Impeller Retaining Nut, which is a plastic insert type.

2. Retaining Washer, which secures the impeller on the shaft.

3. Compressor Impeller, which is an aluminum casting (but is a forging on the 710 model) of blades used to pump air. Here are the numbers of blades you will find on various models:

567 & 645E/EB Models	16 Blades
645EC & 645FB Models	22 Blades
710G Model	34 Blades

4. Impeller Spacer, which is a machined washer that constitutes part of one of the three air seals along the rotating assembly shaft.

5. Compressor Bearing Journal, which is the finished surface on the compressor portion of the shaft corresponding to the compressor bearing.

6. Heat-Dam Washer, which is a larger washer/disc that has bands on the surface; these bands contact the turbine wheel to minimize metal-to-metal contacts, and thus reduce the heat transferred from the turbine wheel to the bearing surface.

7. Compressor Seal, which is a machined surface on the turbine wheel; this surface constitutes part of the second air seal along the rotating assembly shaft.

8. Turbine Wheel, which is the central hub of the rotating assembly; on this wheel sit all of the turbine blades.

9. Turbine Blades, which collect the exhaust gas flow and thus turn the rotating assembly. Here are the numbers of blades you will find on various models:

| 567 & 645 All Models | 47 blades |
| 710-G Model | 53 blades |

10. Sun Gear Shaft, which forms the "rear" end of the rotating assembly (the turbine wheel forms the "front" end); the sun gear shaft has three distinct components
    A. Turbine Seal, which is a machines surface on the shaft that constitutes part of the third air seal along the rotating assembly shaft.
    B. Turbine Bearing Journal, which is the finished surface on the turbine end of the shaft corresponding to the turbine bearing.
    C. Sun Gear, which is part of the sun gear shaft, and serves as the central gear in the planetary gear driven system.

Rotating Assembly

5. Turbocharger Bearings

The rotating assembly is supported by two bearings located in the doweling assembly. Because the turbine wheel experiences high speeds and temperature levels, EMD designed and constructed these two bearings in a rather unusual way.

EMD designed both the compressor journal and turbine journal bearings with cylindrical tapers . The tapers form oil wedges that develop powerful, radially oriented hydraulic forces to center the rotating journals. In other words, instead of using a concentric bore on the inside diameter of the bearings, EMD uses oil "ramps". The hydraulic forces in the EMD journal and thrust bearings far exceed the engine's lube oil pressure. The hydraulic forces are generated by rotating journals, so they increase as rotor speed increases.

On the inner surface of EMD's rotating assembly support bearings, you may discover three, four, or five ramps. Each ramp begins at an oil "channel" or groove. At this point, the clearance between the surface of the bearing and the journal is greatest. But as the ramp winds around the inside of a bearing, its height increases, and the clearance between the bearing and journal decreases. Between the low end and high end of the oil channel, the difference in ramp "heights" is approximately .003 - .004.

How is lubricating oil pumped into the bearings? The rotation of the turbine wheel pumps the oil into the bearings and along the oil ramps. As this oil flows along the ramp, the bearing clearance decreases, which increases the centering force exerted on the journal. (This is known as a "hydrodynamic" bearing design.)

Because silver can carry superior loads, EMD used to silver plate the surface of most of its turbocharger bearings. However, beginning in 1995 EMD switched over to bronze lined turbocharger bearings. All of the other EMD style turbocharger manufacturers still use silver plated bearings. Here are their specific names and functions of the bearings:

1. The Compressor Bearing is the hydrodynamic bearing through which passes the impeller end of the rotating assembly. The compressor bearing supports the compressor portion on the rotating assembly shaft. At its inboard end, the compressor bearing is flared and manufactured with a convex surface to form a part of the thrust bearing assembly. The compressor bearing is located in the compressor bearing support, and is retained by an interference fit.

2. The Thrust Bearing is a disc-shaped bearing through which the turbine wheel shaft also passes. One side of the bearing is concave to correspond with the flared end of the compressor bearing; these curved surfaces thus are self-aligning. The opposite side of the thrust bearing appears flat, but actually consists of a series of tapered pads on the thrust face. These pads form oil wedges that develop hydraulic pressure to separate the bearing from the rotating heat dam thrust washer.

To understand the function on the thrust bearing, you must understand that the pitch of the impeller blades causes the thrust forces found in the rotating assembly. These blades are shaped to pull air through, just like an airplane's propeller pulls air through. Unlike the airplane, a turbocharger impeller must remain stationary in order to pump air through its compressor section. The thrust bearing controls the tendency of the turbine wheel to move forward, a tendency exacerbated by exhaust pressures against the wheel.

The thrust bearing is located between the flared edge of the compressor bearing and the heat dam washer on the turbine wheel.

3. The Turbine Bearing supports the sun gear end of the rotating assembly. Constructed like the compressor bearing (but without a flared edge), the turbine bearing is located in the "clutch support"; the clutch support itself is located in the turbine bearing support.

4 - 6 The Planet Gear Bearings comprise three identical bearings, one for each of the three planetary gears. Unlike other bearings, these bearings have an oil ramp on their outside diameters. You'll find one bearing in the bore of each planet gear; the gears themselves rotate on the stationary bearings.

6. Turbocharged Labyrinth Seals

EMD turbo seals do not contact the turbine wheel shaft. Instead, the seals utilize air pressure to create seals.

How does this system work? Pressurized air from the compressor scroll is ported through a "bleed" air duct into three "labyrinth" seals. The air actually travels through small holes in the Doweling Assembly to the center of each labyrinth seal. This bore of the seal through which the rotating assembly passes, contains several grooves or "labyrinths". The air flows around these grooves, effectively sealing the area.

Labyrinth seals effectively separate lubricating oil from exhaust gases. However, improperly filtered air can deposit dirt within the air passages; this dirt may restrict air flow and reduce the effectiveness of the seals.

There are three labyrinth seals in the EMD turbo:

1. The impeller seal, located directly behind the impeller, prevents oil in the compressor bearing area from being sucked into the compressor air scroll by the spinning impeller.
2. The compressor seal, located between the turbine blades and the compressor bearing, prevents the migration of oil from the compressor bearing into the exhaust section.
3. The turbine seal, located between the turbine blades and the turbine bearing, prevents the migration of oil from the turbine bearing into the exhaust duct.

7. The Turbine Inlet Scroll

The exhaust inlet scroll delivered high energy exhaust gas to the EMD single-stage turbine. The scroll is a welded assembly made from "chrome-moly" plate. EMD forms the plate so that the incoming gas flow is smoothly and evenly distributed with a minimum of turbulence.

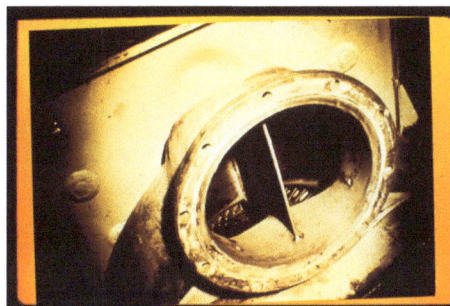

8. The Nozzle Ring

In the exhaust portion or turbine section of the EMD turbocharger, you can discover a series of stationary vanes. The engine's exhaust gases pass through these vanes in order to reach the turbine's blades.

Each passage between the vanes is called a "nozzle". The "nozzle ring", therefore, is simply a series of individual "nozzles" mounted on a common ring.

Since each individual nozzle directs into the turbine wheel blades, the size of the nozzle openings are designed proportionate to the amount of exhaust gas generated by the engine. You will find larger nozzle openings on large engines, and smaller openings on smaller engines. A small nozzle on a large engine might choke the engine's exhaust gas flow. A large nozzle on a small engine would not restrict the gas flow enough to develop the correct velocity as the gas flows through the engine.

Actually, in the EMD turbocharger, the optimum nozzle opening is just enough to permit the maximum engine exhaust gas volume to pass without creating back pressure. Back pressure develops when the gas cannot flow through the nozzle quickly enough and therefore "backs up" into the exhaust system. When back pressure occurs, the turbo "surges", that is, the turbo's gas flow reverses, and the engine actually "burps" exhaust back through the engine!

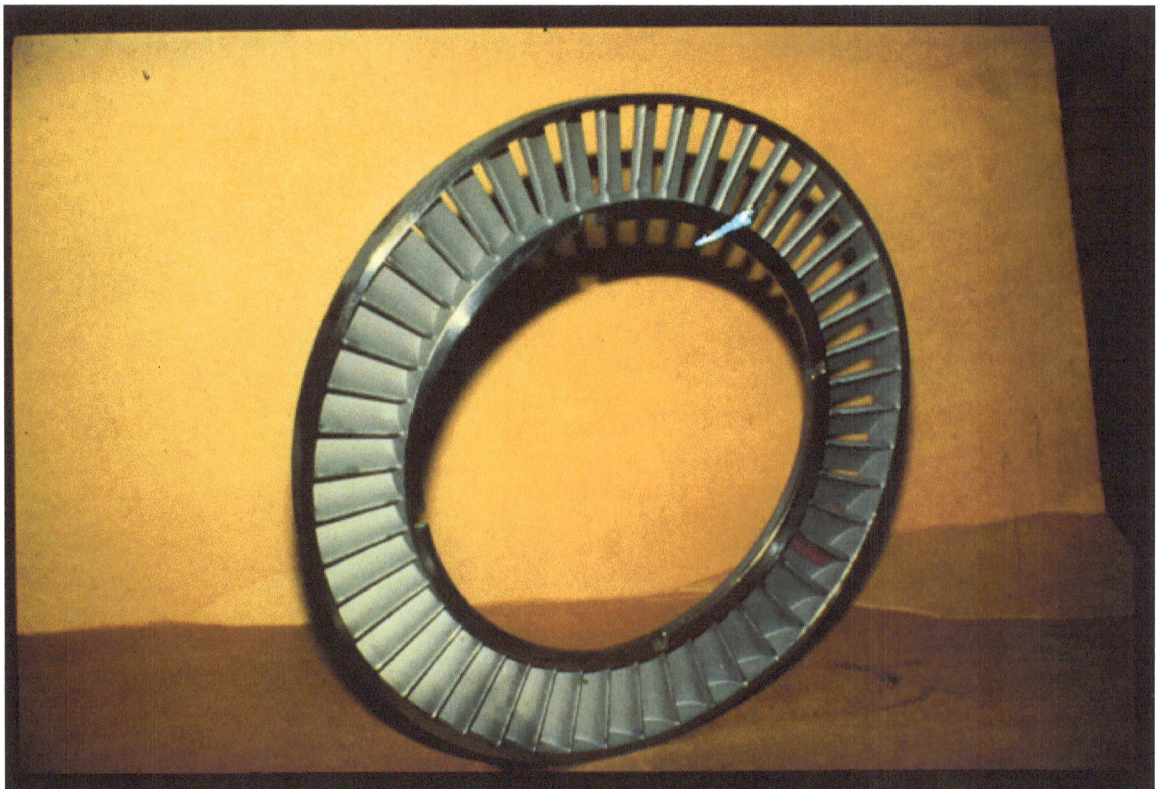

9. The Turbine Shroud and Retaining Clamp

The turbine shroud is a metal band which encircles the turbine blades. By reducing gas leakage around the blade tips, the shroud helps to maximize the exhaust gas flow across the turbine blades. The shroud thus increases the efficiency of the turbocharger.

In designing the shroud, EMD understood that the clearance between the turbine blade tips and the shroud must minimize gas leakage, but still be large enough to prevent the shroud's contacting the blade tips when they enlarge through thermal growth. For this reason, EMD sprayed the inside diameter of its shroud with a "sort metal" abradable coating. In this coating, the turbine blades can establish their own paths as the temperatures normalize. As the blades establish their paths, each shroud becomes "custom fitted" to each blade.

In most turbo models, the shroud is retained by a clamping ring known as the "Marmon Clamp". This clamp consists of 4 channel segments welded to a metal band. The channels engage a flange on the edge of the turbine shroud, thereby securing the shroud within the turbo. EMD also uses a "t-bolt" and nut to apply clamping load to the assembly.

In the early 1980's, EMD improved the Marmon Clamp. In earlier models, EMD spot welded the clamp and the strap to which the t-bolt is attached. Unfortunately, the clamp suffered metal fatigue near these welds after repeated thermal cycling. So EMD began riveting the t-bolt strap to the clamp. This improved clamp has proved more durable in severe service applications.

10. The Exhaust Diffuser

The exhaust diffuser is another aerodynamic device located in the turbine section of the turbo. Basically, the diffuser is an arrangement of 3 or 4 vanes of stationary fins placed directly behind the turbine blades. By controlling turbulence, the diffuser vanes provide a smooth transition path for the exhaust gas as it flows from the turbine blades into the turbo's exhaust duct.

11. The Exhaust Duct, and Eductor Tube

You can find two different exhaust ducts on EMD turbochargers. Until the late 1970's, EMD used a "standard" high duct attached to the main housing by 3 spring washer sets on each mounting foot. Since then, EMD has used the "big-foot" duct, which is two inches shorter than the standard duct and attached to the turbo by a five spring washer set on each mounting foot. The decreased length and increased reinforcement accommodate an exhaust silencer now added atop the turbo duct.

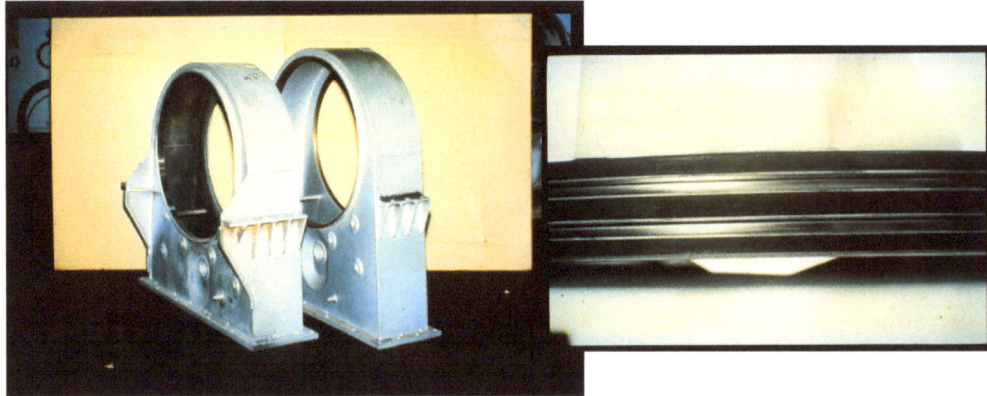

In both types of ducts, EMD built in an aspirator tube that allows for the installation of an "eductor tube". This "eductor tube" produces a suction that ventilates the engine crankcase.

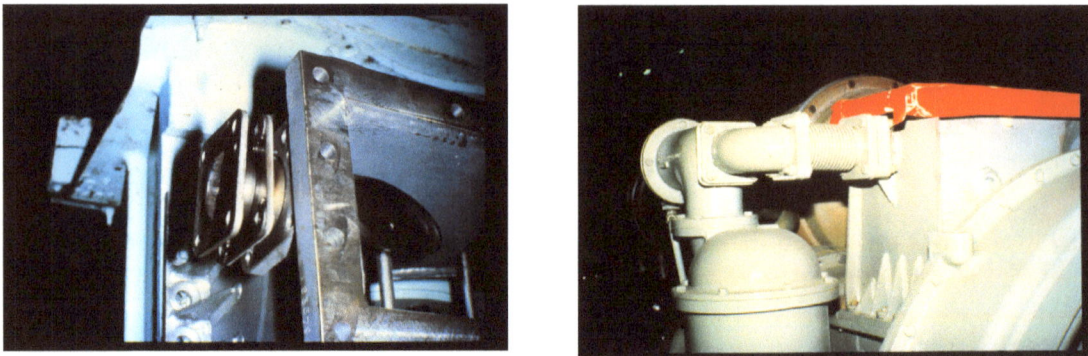

Here's how the "eductor tube" works: as exhaust gas flows upward and out of the duct, a negative pressure occurs behind the beveled end of the tube. The outboard, flanged end of the tube is connected to a screen called the lube oil separator. The suction applied above the screen in the separator draws vapors from the engine's crankcase, while the screen prevents lube oil from being drawn out.

Locomotives, and most marine and industrial, turbochargers with exhaust silencers use an "ejector" system. This system uses (a) compressor discharge air directed through a venturi; and the eductor tube suction, to aspirate the crankcase of these engines. The aspiration is needed because silencers or long runs of exhaust duct work inhibit the exhaust gas flow in these engines.

Near the bottom of its exhaust ducts, EMD provides a drain opening to drain rain water which may enter the duct while the engine is shut down. You can located the drain hole by finding the small tube that passes through the compressor bearing support from the impeller side. The drain hole has a larger diameter than the tube has.

12. The Compressor Diffuser

The compressor diffuser is a row of fins or "vanes" attached to a mounting ring and positioned around the circumference of the impeller. The vanes direct the flow of compressed air discharged from the impeller. Thus, they provide smooth, turbulence-free, high volume air delivery.

EMD manufacturers the compressor diffusers with specifically sized "throat passages" between the vanes. The size of these passages controls the air flow so that the compressor's power requirements are balanced with the power generated in the turbine by the exhaust gas energy at full load. To achieve this control and balance, EMD carefully matches the compressor diffuser throat area to the turbine nozzle ring area during turbo assembly.

13. Planet Gears

Machined onto the end of the turbine wheel, the sun gear meshes with 3 planet gears. These gears engage the sun gear at 120' intervals, and are located by a planetary gear carrier shaft. Basically, the carrier shaft is a disc from which extend three pins. Each pin passes through the center of a planet gear. The opposite side of the carrier shaft disc is splined.

EMD has two basic planet gear designs: the original or "standard" 32 tooth gear, and the "high capacity" 47 tooth gear. The original planet gear performed satisfactorily on railroads, but when EMD turbocharged engines began to be used for high gear-loading application (such as in generating set installations), EMD needed a more durable gear design. All gears transmit varying vibrations as their teeth mesh, and the newer, high capacity design handles vibration better.

To understand this better, consider the example of generating sets, which run at constant high rpm regardless of the load on the system. They must maintain this constant speed to maintain the electrical "line frequency". Yet, their engines often run at less than full rated load, with lower-than-normal exhaust gas energy levels.

In order to compensate for the less powerful exhaust, the turbo's gear drive must work harder. Over time, the result is accelerated planet gear wear.

To solve these potential problems, EMD reduced the vibration level by increasing the tooth contact ratio in the gears. The company also made wider planet gears to distribute loading over a larger area. These "high contact" or "high capacity" gears significantly reduced the light load vibratory levels. They also nearly eliminated tooth wear under high gear train loading conditions.

EMD now uses the high capacity gear train in marine drilling and marine propulsion engines, industrial generator sets, and rail engines.

Note than turbos using the standard gear design carry 18:1 or 19.7:1 gear ratio designations. Turbos with high capacity gears have ratios of 16.7, 16.8, or 17.9:1. These ratios refer to the speed differential between the turbocharger impeller and the engine crankshaft.

14.The Ring Gear and Clutch Housing

In the planetary gear train, the third element is the ring gear. The ring gear surrounds the three planet gears, and is manufactured with teeth on the inside diameter. Consequently, each planet gear's teeth are simultaneously engaged to two gears: (1) the sun gear on the turbine wheel shaft; and (2) the ring gear which surrounds them.

Bolts attach the ring gear to a housing which encases the turbo. These bolts lock the ring gear to the clutch housing.

You may find the clutch housing itself within the turbine bearing support, which is part of the doweling assembly. The clutch housing rotates on the outside diameter of the clutch support (where the turbine bearing is located). Between the clutch support and the clutch housing, EMD used bronze thrust washers and bushings as bearing surfaces.

15. The Clutch Camplate and Rollers

EMD's overrunning clutch design allows rotation in one direction, and engagement or "lockup" in the other direction. To accomplish this, EMD uses a center hub called the support, a set of cylindrical rollers, and a surrounding ring called the camplate

The camplate comprises 12 wedge-shaped pockets with rollers in each pocket. Because EMD designs the 12 pockets with angled ramps in each, the distance varies between the outside diameter of the support and the ramp. More importantly, the pocket depth at one end of the ramp exceeds the diameter of the roller, but at the other end is less than the roller's diameter. Consequently, when it approaches the narrower end, the roller wedges itself between the support and the camplate ramp, thus locking the two parts.

Look at the example illustrated. Notice that if the camplate were rotated in a clockwise direction, the rollers would move into the wider ends of the camplate ramps. The camplate can now rotate free of support. However, if the camplate were turned counterclockwise, the rollers would move into the narrower ends of the ramps. They would lock the camplate to the stationary clutch support.

The camplate is located in the clutch housing. On the end of the clutch housing you may discover the ring gear. Because 6 "drive pins" or dowels attach the clutch housing to the ring gear, the camplate and ring gear actually operate as one unit, though each is attached to the clutch housing at opposite ends.

Since the clutch support is bolted to the turbine bearing support (and thus is stationary), when the camplate locks to the support, the clutch housing and the ring gear also lock.

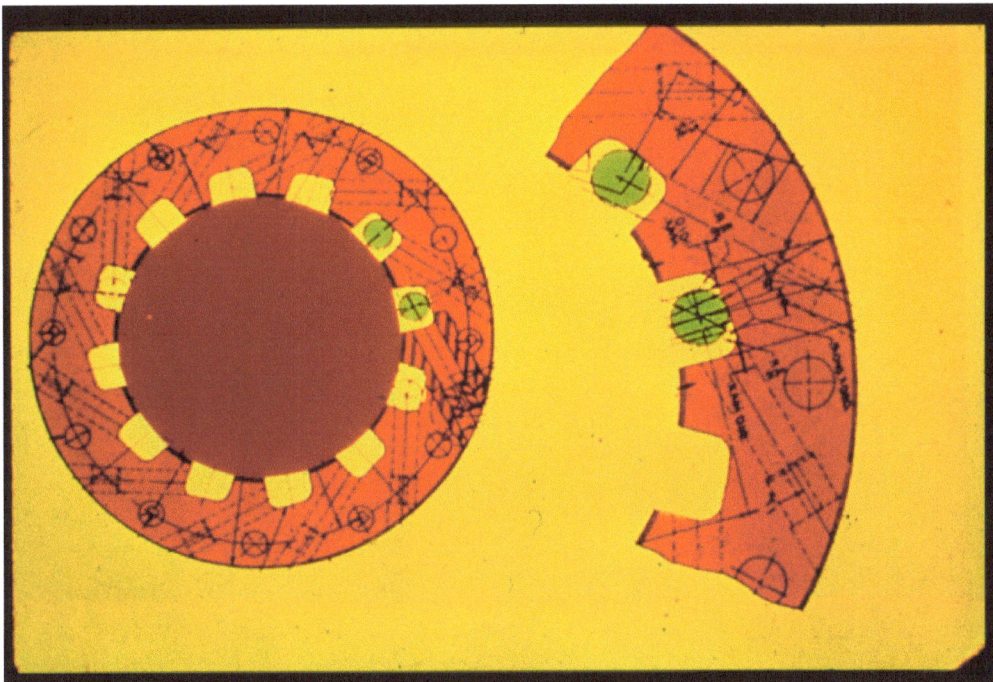

16. The Gear Drive System

The splined end of the carrier shaft extends through the idler gear support.
The idler gear support is the plate at the back of the turbo.

Two bearings support the carrier shaft: (1) a ball bearing located in the idler gear support; and
(2) a roller bearing located in the carrier bearing support.

On the splines of the carrier shaft, EMD mounts a carrier drive gear. Although EMD mounts
it externally on the turbo, you cannot easily see it because the carrier bearing support obscures
your view.

The turbo idler gear is located on a small stub-shaft attached to the idler gear support, below the
carrier gear. The idler gear is mounted on a special, barrel-faced roller bearing. With
barrel-shaped rollers, this bearing self aligns up to 1/8 inch (this is why it's not necessary to
check backlash on the turbo itself). As a result, if you apply force, you can actually wobble the
idler gear on its stub shaft.

The idler gear is engaged with the carrier drive gear at the top, and with the engine's gear train
at the bottom.

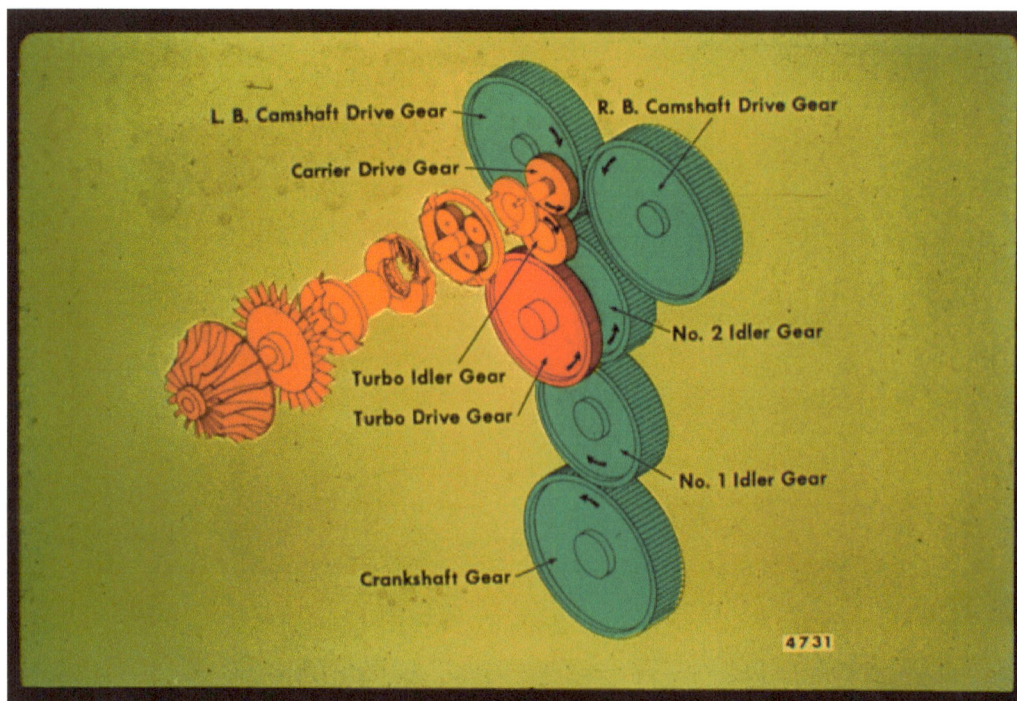

17. The Gear Drive System: Right-Hand Drive Application

In some marine propulsion applications, EMD employs two counter-rotating engines. In such installations, two engines share a common hull, but one has a standard left hand rotation, and one has right hand rotation

Since the gear train of the right hand rotation engine turns opposite the gear train of the standard engine, EMD utilizes a special turbocharger. On the right hand rotation engine, the turbo uses two idler gears rather than the single gear common to most models. These gears permit the turbine wheel to drive in the same direction as all other EMD turbochargers.

18. The Lube Oil System

On the EMD turbocharger, the lubrication system is actually an extension of the engine's oil system. Here is how the oil flows:

1. As oil travels through the main oil gallery in the crankcase towards the rear of the engine, the oil enters the stub shaft bracket assembly on the end sheet of the engine.
2. In this bracket, and oil passage directs the oil to an oil manifold which is also attached to the end sheet. The oil flows through the manifold to the turbo oil filter mounted on the engine.
3. Oil flows through this filter, which carrier the same filters in the main filter tank.
4. If the filter is not plugged, the oil will leave the filter and flow back through the lower leg of the oil manifold. It flows through the small oil pressure sensing line which connected the engine governor to the down stream side of the turbo filter. NOTE: IF the filter IS plugged, no oil will flow and the engine will shut down.
5. Oil flows through another grooved passage in the engine stub shaft bracket, and is admitted to the upper idler stub.
6. Oil flows through a passage in the center of the stub shaft into the 4 inch bore in the turbo's main housing.
7. In the turbo's main housing, a vertical passage (called the "main oil supply") directs the oil upwards.
8. At its top, the main oil passage emits oil for the clutch and planetary gears. From the main passage, oil also passes through a branch line to the auxiliary generator drive; this branch line also delivers oil to the com   pressor and turbine bearing lines.

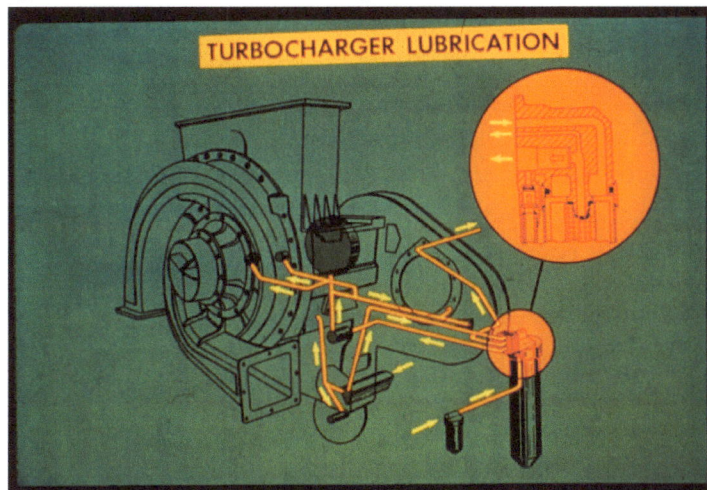

TURBOCHARGER LUBRICATION

19. The Soak Back System

Because the turbocharger depends upon the engine's main lube oil system, EMD adds another lubrication system to protect the turbo when the main oil system flow is unavailable.

To understand the need for this additional system, remember that the main oil system is driven by a gear train connected to the crankshaft. Consequently, oil flows only when the crankshaft turns.

After the engine shuts down, the crankshaft turns another 5-10 seconds. But because of its momentum, the high speed turbine wheel runs for another 30-40 seconds. During most of this time, the main oil system provides no lubrication to the turbo's bearings.

To provide lubrication to the turbo bearings after the engine shuts down, EMD mounts an electrically driven oil pump on the engine. This pump is called the "soak-back pump".

As the EMD engine shut down cycle begins, the soak-back pump supplies the turbo with oil. After the engine stops, the pump operates for another 30-35 minutes. The pump uses the flow of lube oil to carry heat from the turbo's seals and bearings (thus the name "soak-back" pump).

The soak-back pump also energizes during the engine start-up sequence. The pump supplies oil to the turbo bearings even before the main oil system flow can reach them. In this way, the soak-back pump pre-lubricates the bearings. This pump must run three to five minutes before starting the engine!

20. The Planetary System Oil Drainage Screen

From the turbo's planetary system, lube oil drains through openings in the idler gear support. But if the planetary system fails, lube oil may carry off metal fragments and broken gear teeth. To prevent these metal fragments from entering the engine oil sump or passing through the diesel engine's rear gear train, EMD installs a screen in the idler gear support.

Originally, EMD put the screen in a small, triangular-shaped opening in the idler gear support. Most planetary system drain oil flows through this area.

However, with the development of turbos equipped with high capacity type planet gears, EMD needed more oil drainage because high capacity gears have a higher oil flow rate. To these turbos, EMD added three slotted passages on the face of the idler gear support. Drain oil now flows through these three passages and the original triangular opening.

But EMD also needed a larger screen to handle the greater oil flow. So, in the mid-1980's, the company released an improved screen for retrofit in the high capacity turbo's. The screen is installed on the inboard side of the idler gear support. Oil must pass through it as it flows through any of the four possible drainage paths.

Turbos equipped with this new screen do not utilize the original triangular screen.

21. Operation

The EMD turbocharger utilizes a gear-driven system which transmits energy from the engine's crankshaft to the turbine wheel at the sun gear. The turbo uses this planetary gear system when exhaust gas energy levels cannot drive the turbo wheel, such as during engine starting or during light load operation. When exhaust gas energy increases, the turbo decreases its dependency on the gear drive. Eventually, the turbo doesn't need any help from the gear train.

When the turbo no longer needs the gear drive, an overrunning clutch "disengages" the gear drive. Disengagement occurs when the rotating assembly over speeds the driving gear train while the gears remain engaged.

This power "takeoff" originates at the upper idler gear in the engine's camshaft drive gear train. EMD equips this upper idler gear with a shock damping device which uses packs of coil springs to absorb torsional shocks in the engine's gear train. Attached to this damping device is a turbo drive gear. This turbo drive gear serves as the power takeoff for the turbo's gear train. The gear is isolated from the engine's inherent, potentially damaging torsional vibrations.

In the gear train, EMD mounts the next gear on the "rear" of the turbo at the idler gear support. Not surprisingly, EMD names this gear the turbo idler gear.

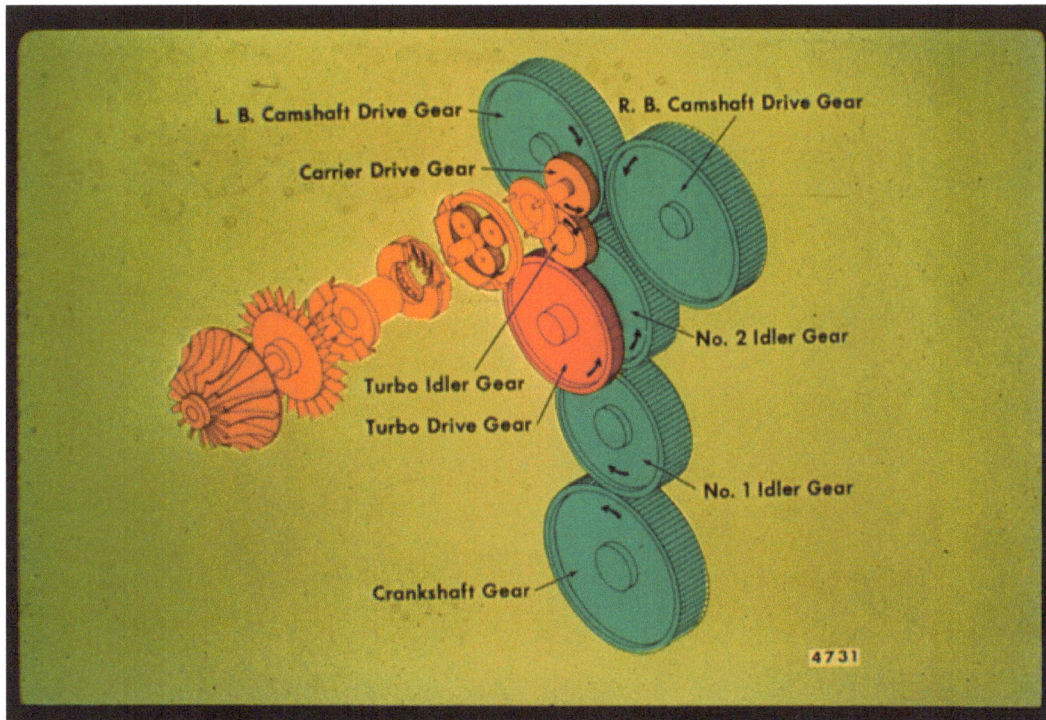

The idler gear drives a turbo-mounted carrier shaft drive gear. EMD locates this gear on the end of the planetary gear carrier shaft. The carrier shaft itself extends through the rear "bulkhead" of the turbo, and carriers the three planetary gears.

22

The planet gears are engaged to both the sun gear on the end of the turbine wheel, and to a ring gear. The three planetary gears surround the central sun gear, and mesh with the sun at 120 degree intervals. A ring gear also surrounds these planetary gears. EMD manufacturers this ring gear with internal teeth, so that the three planetary gears can move along a "track". EMD also attaches the ring gear to the clutch camplate, so that these two components operate as one. If the camplate rotates, so does the ring gear. Conversely, if the camplate locks, the ring gear cannot move.

To understand how the engine gear train drives the turbine wheel, study the following sequence:

1. As the starter motor pinions engage the flywheel, the crankshaft rotates.

2. In the camshaft gear train, the lower idler gear is turned by the force transmitted from the gear teeth on the crank gear to the teeth on the lower idler.

3. The lower idler gear teeth transmit force to the upper idler gear teeth (which they engage). This force turns the upper idler/spring-drive gear assembly.

4. The turbo drive gear assembly (on the spring-drive gear) transmits force to the teeth of the turbo idler gear.

5. The turbo idler gear teeth turn the carrier shaft drive gear.

6. The carrier drive gear turns the entire carrier shaft assembly.

7. Located in the rotating carrier shaft, the three planetary gears pass on torque to both the sun gear and the ring gear.

8. Torque input turns the ring gear (and the clutch housing) in the opposite direction. However, after very little movement the camplate's rollers wedge themselves in the ramps, and the camplate locks to the clutch support.

9. With the ring gear stationary, gear train torque is transmitted through the planet gears to the sun gear. This causes the sun gear to drive the turbine wheel in a counterclockwise direction (as viewed from the impeller). Because of the speed-increasing nature of the planetary gear system, the sun gear rotates at a speed significantly higher than the speed of the carrier shaft assembly that drives it.

10. As the impeller rotates, air drawn through the engine filters increases in velocity, and passes through the compressor diffuser and air scroll with minimum turbulence. In the compressor diffuser, the size of the air passages controls the air flow so that the compressor power requirements are balanced with the power generated in the turbine by exhaust gas energy at full rated load. (For this reason, when a person assembles the turbo, he must "match" the compressor diffuser throat area to the turbine nozzle area.)

11. As air pumps into the engine, combustion begins. As the engine runs, exhaust gases from individual cylinders flow through the turbine section of the turbocharger. The energy in these gases push on the turbine blades; this force helps the engine gear train turn the rotating assembly.

12. The planetary gear system receives two sources of torque. First, turbine torque fed through the sun gear; second, the torque transmitted by the gear train fed through the carrier shaft to the planet gears. The difference between these two torque inputs is the torque transmitted to the ring gear.

13. As we have already discussed, when the turbine does not develop sufficient power to turn the rotor at the engine-determined driving speed, the torque input through the planet gears "locks" the ring gear and camplate. But when the turbine can drive the rotor faster than the speed dictated by the turbo gear ratio, the increased torque from the sun gear feeds through the planet gears to rotate the ring gear and camplate in the "unlocked" direction. The clutch housing now rotates around the clutch support at a speed which equals the difference between the engine gear train speed and the turbine wheel RPM. During this overrunning condition, the clutch rollers move in the wide end of the wedge-shaped pockets formed by the camplate ramps.

14. The turbo operates in this "free-wheeling" state as long as the exhaust gas energy level and flow rate can provide enough power to drive the rotating assembly faster than the gear train would. However, if the engine's speed or load decreases, the amount of exhaust energy decreases, and the turbine's speed begins to drop. When the turbine speed returns to the speed of the gear train, the clutch reengages, and the gear train once again provides some of the energy needed to drive the rotating assembly.

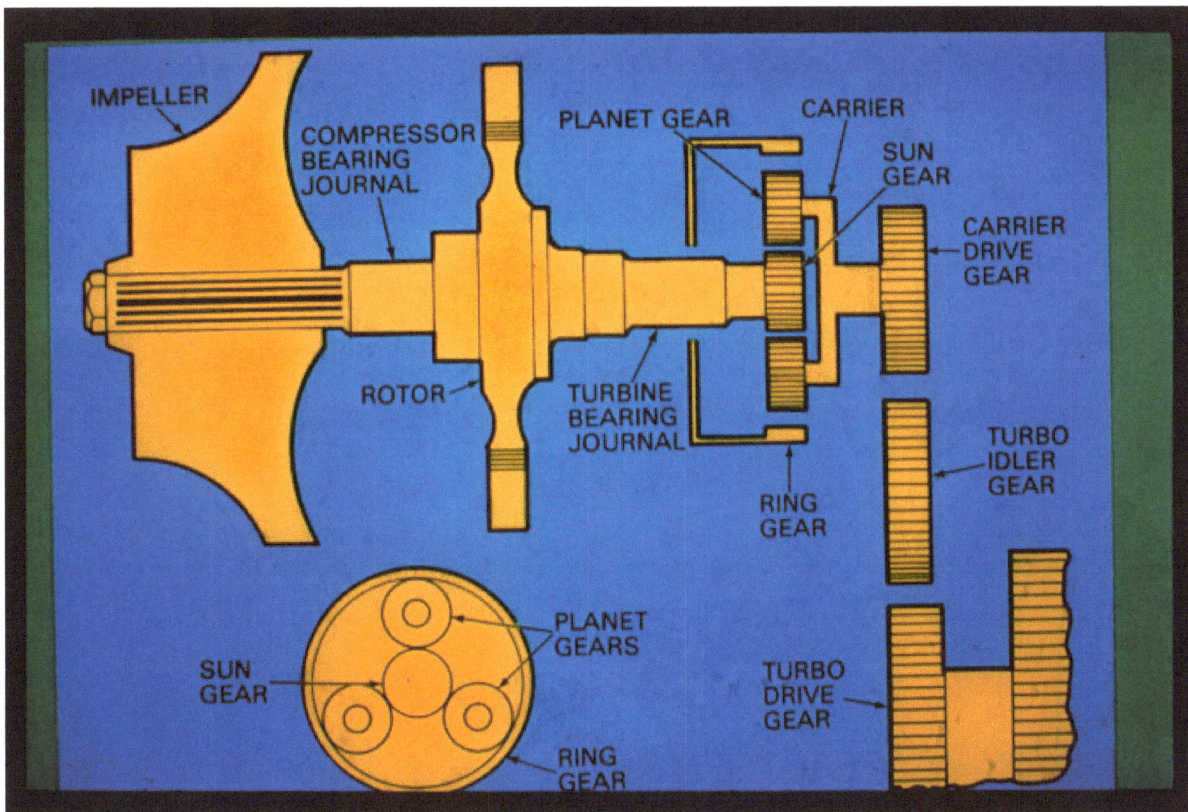

22. Turbochargers with an External Clutch

All 567-T and most 645-E/F turbos utilize the internal 2 roller clutch design which we have already discussed. However, in the early 1980's, EMD field tested an experimental "external clutch". EMD conducted these tests with 645 turbos primarily in marine towing service, as well as other applications with severe loads.

With the 710-G engine, the external clutch became standard equipment. This design removes the clutch from within the turbo, and instead places it in the engine camshaft drive gear train. Instead of the spring-drive gear assembly found in earlier 645 series turbos, the new double gear assembly interconnects by means of a large version of the roller clutch configuration. To be specific: The new clutch utilizes 16 three-quarter inch diameter rollers, in place of the 12 half-inch diameter roller found in the old internal clutch. These new rollers are 1 ½ inch long, in contrast to the 1 inch rollers on the old 12 roller clutch. Likewise, the new external clutch has a camplate diameter of 11.750 inches, versus 7.750 inches in the old clutch. Also, on the new clutch, the camplate roller pockets are inverted or open towards the outside diameter, instead of facing the center of the part, as on the old clutch.

With bigger components and more rollers, the new external EMD clutch can carry much greater loads than the old internal clutch.

Both clutches operate identically. However, in the new external clutch, since the clutch disengages in the engine gear train, the turbo's planetary ring gear now "locks" in the stationary position. This lockup device occupies the space previously occupied by the roller clutch on older models.

In addition, on the new clutch, EMD has added a row of gear teeth to the clutch support. The outside diameter of these teeth equals the diameter of the three planet gears in the carrier shaft assembly.

And on the new clutch, the ring gear is much longer, and has rows of two identical teeth cut on the inside diameter. The new ring gear resembles the kind of sleeve used to synchronize gears in automobile transmissions. The clutch support teeth enter the ring gear at one end, while the planet gears enter from the other end. Because the clutch support is fixed in place, its tooth engagement with the ring gear prevents rotation on the ring gear.

Other manufacturers offer the external clutch for 645 engines. This clutch could be a dramatic improvement over the 12 roller and sprag-type clutches now found in most 645 turbos.

SECTION IV

EXTERNAL INSPECTION AND OPERATIONAL PROBLEM DIAGNOSIS

How can you troubleshoot and qualify your turbos?

If you read engine manuals, troubleshooting guides, or other recent material, you will find some helpful information. But because the EMD turbo has evolved with experience, you may find that past procedures are obsolete or needing revision. Therefore, in Section 4, we offer you some recent revisions, as well as "old" knowledge, about troubleshooting and qualifying your EMD turbos.

Checks You Can Make While Your Turbo Is Still On The Engine

The Roller Clutch Test

1. Idle your engine until it reaches normal operating temperature. (If you cannot start the engine, re move the rubber boot from the turbo inlet, and verify that the impeller locks when you try to turn it in a clockwise direction by hand. If the impeller does not lock, either the clutch has failed completely or the planetary gear train has failed.

2. With the engine warmed up, push inward the injector manual control linkage lever; increase the engine speed to approximately 700 RPM.

3. Pull the injector manual control linkage lever to the "No Fuel" position, overriding the engine's governor. (The clutch will now disengage and allow the turbine to spin free of the gear drive.)

4. As the engine begins to stall, once again push in the injector linkage manual control lever. By providing more fuel, you should increase engine speed. The decelerating turbine wheel will "meet" the accelerating engine gear train, and the roller clutch should engage, providing sufficient air for continued engine speed increase.

5. If the clutch fails to engage, the injector rack linkage will move towards the "full fuel" position, black smoke will emit from the exhaust duct, and the engine may stall. These symptoms indicate imminent clutch failure. You should replace the turbocharger.

Turbocharger roller-type clutches tend to fail gradually, rather than suddenly. In the early stages of clutch wear out, the slippage may be intermittent. In such instances, the engine may smoke heavily or stall during speed changes, yet behave normally later. If you want to be SURE that the clutch has not reached this early stage of failure, you may repeat steps 1-5 a few times. But contrary to some reports, you do NOT need to make 30 consecutive tests! In fact, to avoid damaging a good clutch, you should not perform injector linkage manipulation more than 2 or 3 times. Within 2 or 3 tests, a defective clutch will exhibit failure symptoms.

Since 1976, EMD has built almost all 567 and 645 turbos with a roller-type clutch. NEVER USE A TORQUE WRENCH ON THE IMPELLER NUT!

The Turbocharger Oil Pressure Test

Sometimes you might want to confirm that the main engine and soak-back oil systems are actually delivering lube oil to the turbo. For example, you would want to conduct this test after a manufacturer had installed a turbo but not run it on a test cell. Or you might conduct this test after you had replaced a turbo because of a :bearing failure". Remember that turbocharger bearing failures are usually caused by external factors such as imbalance of the rotating assembly (due to foreign object damage) or a lack of proper lubrication. Therefore, when you observe an impeller rubbing the inside of the turbo's air inlet portion without damaging the turbine blades, you should confirm the flow of lube oil through the new turbo before you return it to service.

1. Locate and remove the compressor bearing oil passage pipe plug on the right bank side of the engine's turbocharger. EMD installs this plug in the compressor bearing support, the 3 inch thick casting between the main housing and the air scroll. The ½ inch Pipe plug accepts a 3/8 inch male square drive like the kind on an ordinary ratchet wrench. You can find the plug above the right-bank air scroll to after cooler duct mounting flange.

2. Temporarily install an oil pressure gauge to the oil passage.

3. While you look at the oil pressure gauge, operate the soak-back pump. The gauge should indicate oil pressure in the 15-30 PSI range. If you read no oil pressure, then do not start the engine until you determine the cause. (By the way, if oil does flow, check that no oil is flowing from the engine camshaft bearings. If oil is flowing from the bearings, you probably have a defective check valve in the turbo filter assembly.)

4. Start the engine, and run at a minimum speed (150-200 RPM). Hold back the injector control level until oil pressure builds. Gradually release the lever to prevent engine speed surge.

5. Run the engine at normal idle speed (315-350 RPM) for at least 15 minutes. At the same time, check for;
      A. Constant lube oil pressure at the turbo bearings (use the gauge attached to the compressor bearing support);
      B. Abnormal gear or rotor noises;
      C. Air leaks at the compressor discharge connections'
      D. Exhaust leaks between the manifold, screen and turbo inlet.

6. Increase the engine's speed to 475-483 RPM (no load) for about 10 minutes. The oil temperature should reach approximately 100'F. No recheck items a, b, c, and d mentioned above.

7. Increase the engine's speed to approximately 650 RPM (no load), and run it for fifteen minutes, or until the oil temperature reaches 130'F. By reading the oil inlet pressure gauge mounted on the accessory rack (which is connected to the low oil pressure sensing line in the governor) and by reading the temporary gauge in the compressor bearing support, you can monitor the pressure differential, through the turbocharger. At 650 RPM, the oil pressure difference should range between 5 and 34 PSI. If the difference exceeds 34 PSI, you should determine the cause of this high flow rate.

8. Remove the oil pressure gauge, and re-apply the pipe plug. CAUTION: Do not overtighten the plug! Its casting is easily cracked

The Run-Down Test

1. Allow the engine to run at idle speed until its temperatures normalize.

2. If possible, apply load to the engine.

3. Run the engine at full speed (with full load, if possible). CAUTION: Do not run the engine for more than 3 minutes at full speed/no load. If you do, you will put an abnormally heavy load on the planetary gear train.

4. Remove load from the engine.

5. Abruptly stop the engine from full speed. Use the low oil pressure shutdown, emergency fuel cutoff, overspeed trip, or a similar method.

Additional Turbocharger External Inspections

Ponder this: 60-75% of all "turbocharger failures" are caused by external sources, such as foreign object damage, overheat/over speed, lack of proper lubrication, etc. To avoid these kinds of failures, and to ensure that these failures do not recur, you must perform ALL diagnostic inspections. If you observe a problem, do NOT stop inspecting. In most cases, you must observe and review SEVERAL symptoms to fully understand and correct your problem.

EMD turbos have four external inspection areas:

1. Air Inlet and Impeller Inspection

Remove the air inlet boot, and inspect for the following conditions:

A. Broken impeller blades, which may indicate foreign object damage or metal fatigue;

B. Nicked leading edges on the impeller blades, which may indicated foreign object passage through the air stream. Check the air filter box, the air duct, and replace the air filters;

C. Blade rub on the inside of the cast iron impeller cover, which may indicate the loss of turbine wheel support caused by compressor, turbine or thrust bearing failure. You MUST determine the cause of the bearing failure. Note, too, you should always replace air filters, and check after cooler cores, after cooler ducts and the air box for aluminum debris;

D. Impeller lock up, which you test by hand turning the impeller counterclockwise. Counter clockwise, the impeller should freely wheel; clockwise, it should "lock up". If the impeller free wheels in both directions, either the clutch has failed completely, or a planetary system has failed. If the impeller cannot turn in either direction, then the rotor is locked up, and the inspection should continue.

2. Exhaust Outlet Inspection

Look down the exhaust duct of the turbo, and inspect for the following conditions:

A. Warped exhaust diffuser, which you can identify by "wavy" looking exhaust diffuser vanes. These vanes warp when the turbo experiences an "overheat-overspeed" condition. The turbo's diffuser becomes overheated, and the heat causes its thin metal vanes to distort permanently. A warped diffuser always indicates excessive engine and exhaust gas temperatures.

You must look for causes of excessive heat. Remember that excessive heat causes the turbine to spin faster than it normally does (hence the name "overheat-overspeed"). As the turbine spins faster, its blades being to soften and stretch, and they may eventually break. Also, the impeller tries to pull the turbine wheel forward out through the air inlet. This pull overloads the thrust bearing, which eventually fails itself.

What can cause excessive heat energy? Broken piston rings, worn injector tips, broken valves, improperly timed fuel injectors, incorrect valve timing, plugged aftercooler cores, and plugged engine filters are all possible causes.

Any of these conditions can also provoke either an "air box fire" or an "exhaust manifold fire". You can find evidence of such fires in the form on gray-colored ash near the fire's location. Thus, when an over-heat-overspeed failure claims a turbo, you must inspect the air box and the exhaust manifold with a bright lamp. If you don't identify the cause of the fires, the same problem will damage replacement turbos. Note, to, that when thick, wet, sponge-like soot deposits on an air box reach ½ inch in depth, you should clean the air box.

B. A bulged or punctured turbine shroud, which results from an overheat-overspeed that causes the blades to stretch and contact the shroud surrounding them. (Recall the small clearance between the blades and the shroud.) The blades may deform, bulge, or even puncture the shroud.

C. A broken shroud retainer clamp, which often results from metal fatigue. Since this clamp secures the shroud in most turbos, if the clamp breaks, the shroud may drop and damage the turbine blades. Therefore, if you observe a broken shroud retainer clamp, remove the turbo immediately.

D. Oil out the exhaust stack, which may result from (1) insufficient air reaching the turbo's seals, perhaps because clogged air filters are restricting needed air inflow to the turbo seals (check the filter pressure drop before you change the whole turbo!); (2) oil coming from the engine, not the turbo (again, before you change the whole turbo, remove the exhaust screen and check the turbine inlet for wet, shiny deposits which indicate engine, not turbo, oil).

3. Exhaust Inlet Inspection

Remove the turbo's inlet screen, and look for the following possible conditions:

A. Wet, oily deposits, which indicate the engine has an oil control problem that might ignite an exhaust manifold or air box fire. If the engine is OK, the inlet will appear dry and have a flat black, sooty coloring.

B. Bent or plugged nozzle ring passages, which you can observe with a bright lamp aimed into the turbine exhaust inlet. As you look, you will see the stationary vanes that constitute the nozzle ring. Through this ring gas must flow to the rotating assembly. If you see a dented or bent nozzle, a foreign object has probably passed through it. And if you see any deposits on the nozzle's opening, the engine may have a water leak (though fuel can also leave deposits). In either case, dented or lugged nozzles restrict the turbo's gas flow, and thus can cause the turbo to surge or "burp" at higher engine speeds.

C. Nicked or broken turbine blades, which indicate that foreign material - usually pieces of broken piston rings or exhaust valves - has passed through them. Since these blades cause the rotating assembly to spin whenever exhaust gas flows through them, nicked blades will imbalance the high-speed rotating assembly, and a compressor or turbine bearing may fail if the turbine continues operating. Note that turbine blades may also fail because of severe impact, stretching caused by excessive heat or speed, or metal fatigues. In such instances, the tremendous rotor imbalance will effect a severe bearing failure.

4. External Gears and Oil Drainage Inspection

To carry out this inspection, you must remove the turbo. Then, look for the following:

A. A damaged turbo idler or damaged carrier drive gears, which generally indicate an engine gear train problem, NOT a turbo malfunction. If the turbo rotating assembly has seized, one of the externally mounted gears may exhibit broken teeth.

B. Metallic debris in the oil drainage screen, a small triangular-shaped screen which you can locate just below the turbo-mounted idler gear (Note: in 1988, EMD replaced this triangular screen with a larger internal screen). As it drains from the turbo, all lubricating oil passes from the planetary gear train through this screen. Thus, if a planetary gear train component breaks, the oil will carry this debris and deposit it against the inside of the screen.

Additional Troubleshooting Information

1. Oil Reported Out The Stack

A. Check the engine air filters for plugging. If the turbo's labyrinth seals lack air, oil will migrate across the seals especially at high speeds.

B. Remove the oil eductor tube, and inspect the lube oil separator assembly. Check for a damaged or missing screen, which would permit crankcase vapors to draw out oil.

C. Remove the expansion joint between the turbo exhaust inlet and the exhaust screen assembly. Inspect the turbine inlet scroll. If the scroll is coated with wet, shiny oil, the engine is blowing oil out the stack. Using the Maintenance Manual, inspect the engine and determine the source of the oil.

D. If the turbine inlet is dry, the turbo may have a true seal problem. Since labyrinth seals are non-wearing components, either dirt has plugged them, or the rotor has damaged them from contact caused by a bearing failure. You must remove the turbo.

2. Exhaust Leaks

You can usually discover exhaust leaks at the expansion joints between the exhaust manifold sections, or at the connection to the turbine inlet scroll. These leaks endanger operating personnel and detract from the turbo's efficiency.

If you discover a crack in the turbine inlet scroll, you cannot successfully repair it in place. You must remove the turbo.

3. Noises

Generally, you should not use noise to judge a turbo's condition. Because of manufacturing toler-ances and operating characteristics, identical turbos can make different sounds. Turbos commonly emit "chirping" noises, particularly at low speeds, such as idle. When you release the injector control level from higher engine speeds, you may also hear a "chirp" in varying cadence. You'll hear this noise because the turbine is returning to the gear train.

EXCEPTIONS: if you hear (a) loud screeching noises or (b) severe humming accompanied by vibration, you may have a problem. Check the impeller cover for rubbing marks.

4. Burping and Smoking

Burning and smoking indicate a reversal of the normal exhaust gas flow through the engine and turbocharger. A typical engine has an approximate 2 psi drop across its power assembly.

This means that air box pressure exceeds the exhaust manifold pressure by 2 psi throughout the speed range. But if this condition changes, and exhaust pressure momentarily exceeds air box pressure, you will hear a burp. The burp relieves the excess pressure through the turbo air inlet. The engine's opera-tion may now return to normal until back-pressure again builds.

Turbocharger Gas Flow Reversal

Engine surges can harm a turbo. First, the hot exhaust gas reversal flowing into the engine air box may ignite any combustible deposits in the air box. An air box fire can result in turbo overheat - over speed. Second, surging imposes a significant load upon the clutch and planetary gear drive system.

When you notice surging, you should proceed as follows:

A. Locate another engine of the same type and model, e.g., 16-645E3B.
B. Install a 0-30 PIS pressure gauge on a modified hand hole cover of each engine.
C. Run each engine at full-speed, no load. Record the pressure reading.
D. Since each engine was operating on the gear train (due to no load), each turbo was operating as a gear driven blower. You may, therefore, expect that the two engines' pressure readings will not vary by more than 1 PSI.

6. High Air Box Pressure on Suspect Engine

A. Check the turbine exhaust inlet screen for plugging.
B. Inspect turbine nozzles for plugging; inspect turbine blades for damage.
C. Check cylinder lines inlet ports for plugging.
D. Check valve timing; is it late?

7. Overheat - Over speed Failure

When the turbine wheel over speeds because of excessive exhaust gas temperatures, your engine may experience an overheat - over speed failure. his failure typically destroys a turbocharger.

Symptoms of overheat - over speed include a warped or deformed exhaust diffuser (from excessive temperatures) or stretched or elongated turbine blades (from excessive heat and centrifugal forces acting together) Clutch Failure

Because the clutch has metal-to-metal contact during operation, you must expect it to wear out. EMD currently recommends that to prevent clutch failure, you should replace the turbo every 4 years or 24,000 hours of operation.

A turbo's clutch life can be severely shortened by:

1. Abnormal vibration levels caused by an unbalanced rotor;
2. Worn planetary gears;
3. Frequent abnormal cycling such as from surging;
4. Contaminated lubricating oil.

To understand overheat - over speed, remember that normally, exhaust gas temperatures range from 850 to 1050'F, (full speed, full load); generally, temperatures will vary depending upon the weather, fuel, engine load, etc. External conditions - such as worn power assemblies and dirty air boxes - can bring about overheat - over speed failure. If you do not correct the cause of the failure, your replacement turbo may experience a similar failure.

Indeed, preventive maintenance is all-important! When a turbo experiences overheat - overspeed, the turbine speed may increase at a rate of 5000 RPM per second . If the turbo over speeds while operating near its peak RPM (usually 18,500 to 21,500), within one second the turbo's speed can exceed its safe limits, with severe damage to the turbo. You have no quick way of counteracting the problem; thus, you must act preventively.

To prevent overheat - overspeed, remember that any of the following conditions can increase air box temperature and ultimately bring about overhear - overspeed failure:

1 .Dirty after coolers
2 Broken compression rings
3. Late injector timing
4. Incorrect valve timing
5. Plugged exhaust screen
6. Plugged engine air filter
7. Damaged injector tips
8.An exhaust manifold fire

Foreign Material Damage To Turbine

In the power assembly or exhaust system, the breaking of any part can result in foreign material damage tot he turbine nozzle ring and turbine wheel blades (see figure 4-5). Common sources of foreign material include broken compression rings (resulting from too much land clearance) and fragments of exhaust valves (resulting from improper lash adjustment, failed hydraulic lash units, or excessive temperature).

The exhaust screen reasonably protects the turbocharger from this material. But small objects can still pass through it, and larger sharp objects can eventually tear the screen grid and pass through to the turbo. You must, therefore, periodically remove, clean, and inspect the exhaust screen. If you observe any foreign object damage, you must identify the source and correct the problem, or else risk turbocharger failure.

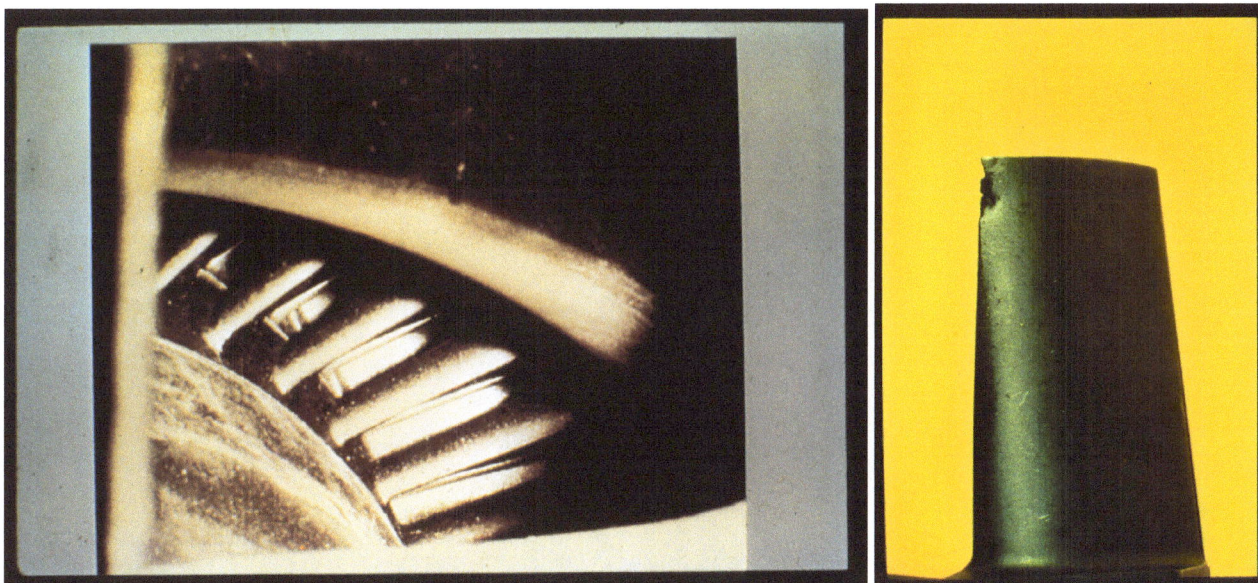

Please note, too, that because the turbine wheel rotates at a high speed, small nicks near the outside diameter of the turbine blades can cause serious imbalance of the rotating assembly. This imbalance damages the turbo bearings and will bring about their failure. SO. . . . . if you observe foreign material damage in the turbine section, remove and repair the turbo prior to running it to destruction. You will save yourself a high repair bill in the future.

Foreign Material Damage To Compressor Impeller

Since a highly effective air filter protects the air inlet, you might expect that few turbos experience damage to their compressor impellers. But, surprisingly, this king of damage does happen frequently enough!

Typically, impeller nicks result from either (a) a previous turbocharger failure, whereby broken pieces from a previous compressor impeller were driven into the air filter or filter housing' and not removed; or (b) a miss-application of the compressor inlet boot or clamp. You must tighten the clamp squarely on the inlet, or it may vibrate loose and enter the turbo.

If you observe nicks on the leading edges of the compressor impeller blades, remember that this damage can result in the same serious imbalance and destruction as that found on nicked turbine blades. You must, therefore, replace the turbo.

Clutch Failure

Because the clutch has metal-to-metal contact during operation, you must expect it to wear out. EMD currently recommends that to prevent clutch failure, you should replace the turbo every 4 years or 24,000 hours of operation.

A turbo's clutch life can be severely shortened by:

1. Abnormal vibration levels caused by an unbalanced rotor;
2. Worn planetary gears;
3. Frequent abnormal cycling such as from surging;
4. Contaminated lubricating oil.

Lack of Proper Lubrication

Improper lubrication has many possible causes. First, the lubrication system may have malfunctioned. A soak-back pump or oil pump may have failed; a lubricating passage may be blocked; or the oil may have become contaminated. But you may also discover that the soak-back oiling was interrupted prior to the completion of its timed cycle. Or you may find that excessive engine RPM during start up has "wiped" a bearing, a result of not manually controlling the injector control linkage to avoid excessive speed during cold weather start-ups. Additionally, you must check the four critical soak back check valves once per year, two valves are located in the turbo filter head, and two in the soak back filter head. Failure of any of these check valves will cause turbocharger failure, in some cases immediate failure!

When a turbo experiences a lubrication failure, most, if not all, of the six silver turbo bearings can exhibit distress such as smearing. Since the compressor thrust bearing is usually the most heavily loaded surface in the turbo, its damage will usually exceed the damage done to other areas such as planet bearing surfaces.

Before you install a replacement turbo, you must correct the cause of lubrication failure. And when you install a replacement, you should use the test described on page 38 to confirm that oil pressure is actually reaching the turbo. You must observe oil pressure during the soak-back pump's operation prior to engine start-up, or the replacement turbo will fail.

Bearing Failures

Turbocharger bearings rarely fail without an external cause. They generally fail because of rotor imbalance resulting from foreign material damage, overheat - overspeed, or insufficient lubrication. By distorting the doweling assembly, misapplied after cooler ducts can also cause bearing failures.

If your turbo experiences a bearing failure, you must inspect it thoroughly to identify all of the related failures. Once you've done this, you can reconstruct the series of events that led to failure in order to deduce the root cause of the bearing failure.

For example, if the impeller has rubbed the cover, the turbo has obviously had a bearing failure. However, upon further inspection, you may discover nicked turbine or impeller blades, caused by foreign object passage that led to the rotor imbalance. You might conclude that foreign material struck the rotor and thus caused a vibration. As the turbo vibrated unchecked, the oil film on the bearing broke down, and the bearing "smeared". The smeared bearing eventually failed to support the turbine wheel, so the impeller rubbed the inside of its cover. Unless you identify and correct the source of the foreign material, your replacement turbo may fail in a similar manner.

Smeared Bearings

Sometimes, just by looking, you can tell if a bearing has failed or suffered severe distress. You must look for the heaviest concentration of aluminum particles on the inside of the impeller cover. Look, too, for impeller blades rubbed on their edges, representing 360 degrees of damage to the impeller.

The impeller cover can also signal major bearing distress. Are aluminum particles distributed evenly 360 degrees around the inside of the cover. If so, assume that the thrust bearing failed, and the rotor moved forward. But if you primarily see aluminum in the bottom or 6:00 position of the cover, assume that the compressor bearing failed, and the end of the rotor dropped. Conversely, if you see aluminum concentrated at the top of 12:00 position, the turbine bearing probably failed, and allowed the sun gear end of the rotor to drop, raising the impeller end.

In any event, you must continue the diagnosis to determine the cause of the bearing failure.

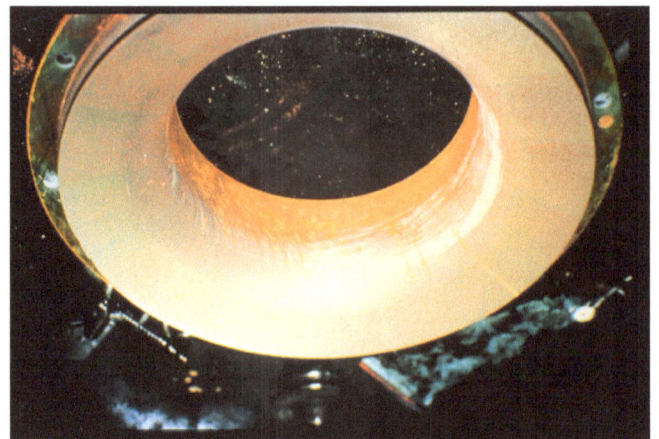

15. Turbine Blade Fatigue

Through metal fatigue, turbine blades can break off the rotating assembly. Although uncommon, blade breakage can damage the turbo considerably. With so much mass removed from one side of the wheel, the extreme imbalance can actually bend the rotating assembly ahead of the compressor journal, while swinging the impeller out towards the "light" side of the wheel.

As mentioned above, a vibrating planetary gear system mesh can cause turbine blade fatigue. Fatigue can result from a manufacturing defect of the blade itself. However, you will generally notice manufacturing defects early in a turbo's life, whereas you will notice gear mesh problems much later in a turbo's life.

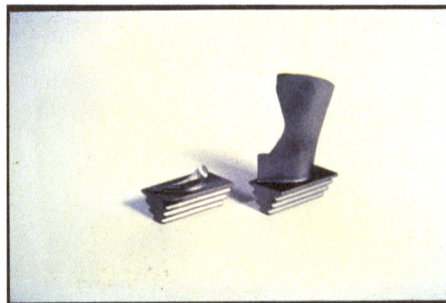

Installation

Before removing the new turbo from the box, chalk mark one impeller blade at the 12 o'clock position. Use a feeler gauge to measure and record the "Impeller Eye Clearance" at 12 o'clock and 6 o'clock, and 3 o'clock and 9 o'clock.

After hanging the turbo, snug the Large bolts, then snug the small bolts

Install Aux Gen Drive assembly. Check backlash, new = .010" - .022", Maximum = .025"
Shift the turbo as necessary to achieve the correct backlash!

Torque the 6 large bolts to 175 ft lbs, start at the center and alternate side to side. Torque the smaller bolts to 65 ft lbs, once again start in the center working outward

Install the right bank aftercooler duct as follows:

Snug the bolts at the Turbo end of the ducts

Torque the bolts at the engine end of the duct to 65 ft lbs

Remove the bolts from the Turbo end of the duct, install the gasket and confirm that an .008" feeler gauge will no enter)

If the feeler gauge does enter loosen and reposition the ducts on the engine. (If necessary
        enlarge the holes on the engine end of the ducts to give yourself more movement)

Torque the engine bolts and verify that a .001" feeler gauge will not enter

Verify that the "IMPELLER EYE CLEARANCE" readings have not changed

If the eye clearance readings have changed, the aftercooler duct must be repositioned!

Repeat this process with the remaining aftercooler duct!

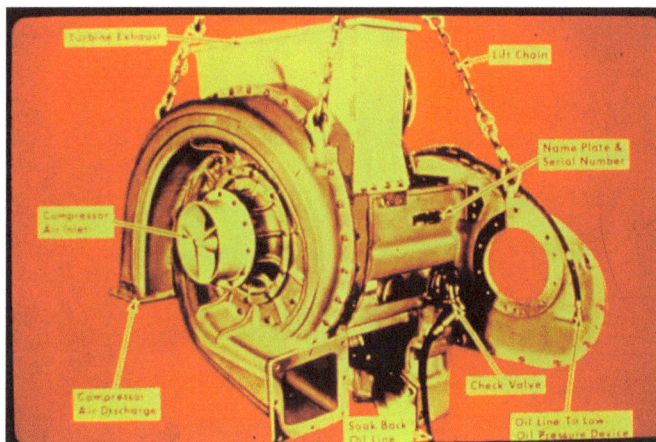

www.ingramcontent.com/pod-product-compliance
Lightning Source LLC
LaVergne TN
LVHW072108070426
835509LV00002B/80